The Thinking Tree

Wild Wildernes

ADVENTURE
HANDBOOK

A Survival Guide
& Science Journal

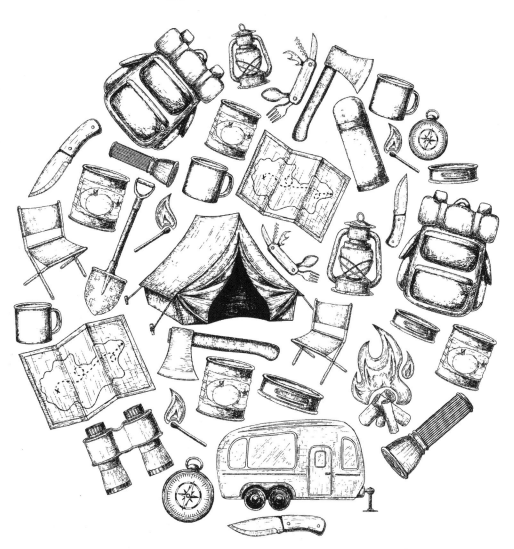

Aiden Potter and Anna Kidalova
Twins for the Wins!

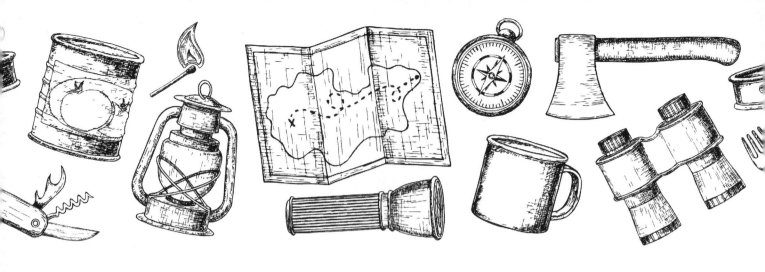

The Thinking Tree, LLC
Creators of Dyslexia Games and
other Incredible Homeschooling Resources

By Aiden Potter, Anna Kidalova,
Sarah Janisse Brown & Andrew McNoodles

Copyright Information

317.622.8852 PHONE (Dial +1 outside of the USA) 267.712.7889 FAX

FunSchoolingBooks.com

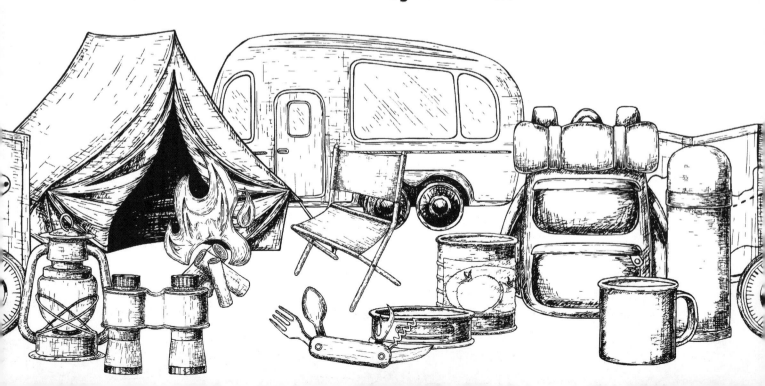

Table of Contents

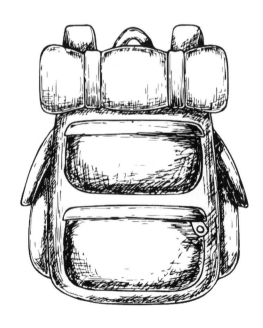

CHAPTER ONE: WILDLIFE

Section 1: Flora

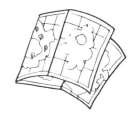

Poisonous plants:

Find information on these poisonous plants and other poisonous plants in your area.

Name of plant:	How it uses poison:	Picture or Drawing:	Other facts:
Poison Ivy			
Poison Oak			

Poison Sumac

Stinging Nettle

Color this poisonous plant correctly.

Poisonous plants are dangerous, but some the are very pretty. Draw your favorite poisonous plant below.

Edible plants:

Find the following information about these edible plants and other edible plants in your area!

Name of plant:	How to prepare it:	Picture or Drawing:	Other facts:
Plantain (the grass, not the fruit)	Eat raw or boiled		
Dandelion	Eat raw or boiled		
Burdock	Boil roots		

Persimmons

8

Color this edible plant correctly.

Many edible plants are a good reminder of God's care. Draw your favorite edible plant below.

Taste Test!

Taste some of the edible plants and record what you think of them below.

Plant:	Reaction:

Medicinal plants:

Find information on these plants that can be used as natural healers and research others in your area.

Name of plant:	Medical Use:	Picture or Drawing:	Other Facts:
Jewelweed	Neutralizer for poisonous plants, itching		
Oregon Grape	Upset stomach, tonic.		
Yarrow	Cure fever and cold		

Color this medicinal plant correctly.

Draw your favorite medicinal plant here.

Section II: Fauna

Dangerous Animals:

Research three dangerous animals from each category that live near your area and answer the questions below.

Reptiles & Amphibians:

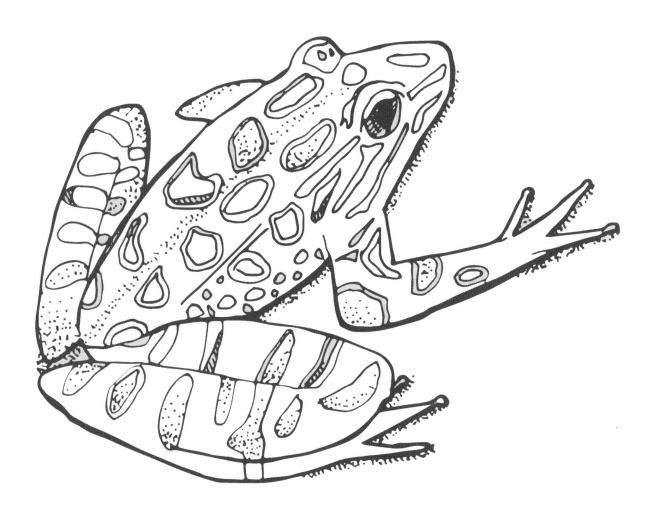

Animal #1:_____

Danger(s) this animal presents:

How to identify this animal:

Tips for avoiding or dealing with this animal:

Animal #2:_____

Danger(s) this animal presents:

_____

_____

_____

_____

_____

_____

_____

How to identify this animal:

_____

_____

_____

_____

_____

_____

_____

Tips for avoiding or dealing with this animal:

_____

_____

_____

_____

_____

_____

_____

Animal #3:_____

Danger(s) this animal presents:

How to identify this animal:

Tips for avoiding or dealing with this animal:

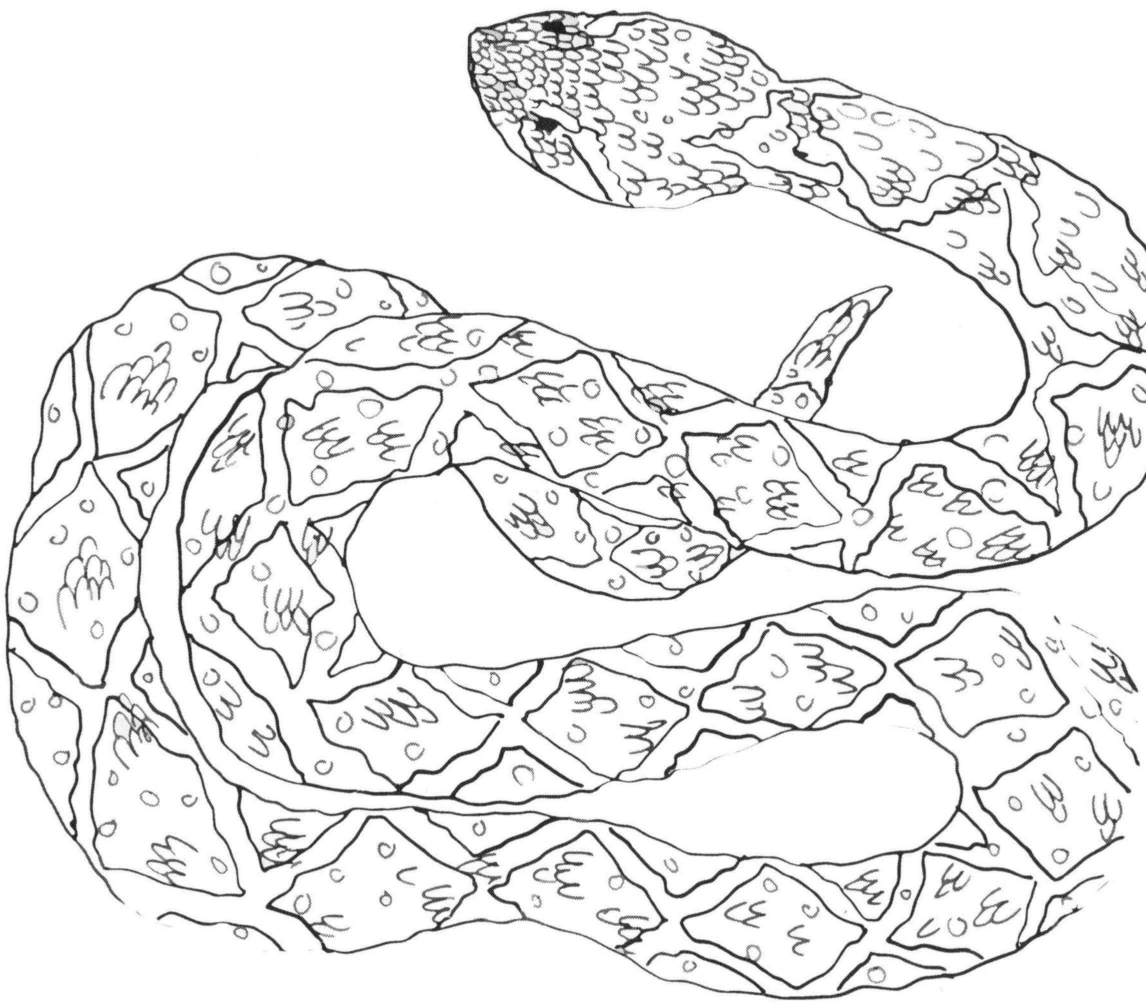

_____

_____

_____

_____

_____

_____

_____

Birds:

Animal #1:_____

Danger(s) this animal presents:

_____

_____

_____

_____

_____

_____

_____

How to identify this animal:

_____

_____

_____

_____

_____

_____

_____

Tips for avoiding or dealing with this animal:

_____

_____

_____

_____

_____

_____

_____

Animal #2:_____

Danger(s) this animal presents:

_____

_____

_____

_____

_____

_____

How to identify this animal:

_____

_____

_____

_____

_____

_____

Tips for avoiding or dealing with this animal:

Animal #3:_____

Danger(s) this animal presents:

How to identify this animal:

Tips for avoiding or dealing with this animal:

_____

_____

_____

_____

_____

_____

_____

Mammals:

Animal #1:_____

Danger(s) this animal presents:

_____

_____

_____

_____

_____

_____

How to identify this animal:

_____

_____

_____

_____

_____

_____

Tips for avoiding or dealing with this animal:

_____

_____

_____

_____

_____

_____

_____

Animal #2:_____

Danger(s) this animal presents:

How to identify this animal:

Tips for avoiding or dealing with this animal:

_____

_____

_____

_____

_____

_____

_____

Animal #3:_____

Danger(s) this animal presents:

How to identify this animal:

Tips for avoiding or dealing with this animal:

Fish:

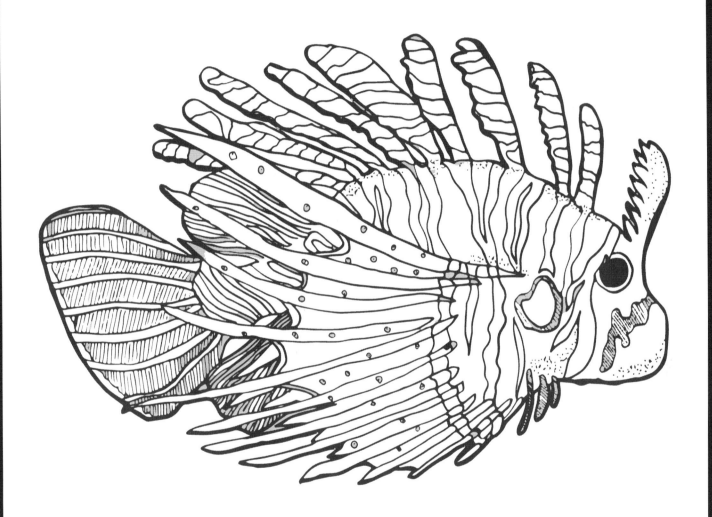

Animal #1:_____

Danger(s) this animal presents:

How to identify this animal:

Tips for avoiding or dealing with this animal:

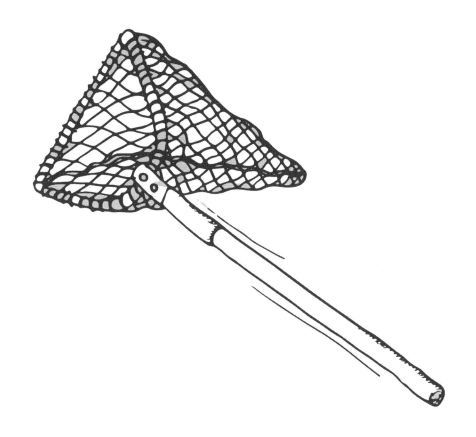

Animal #2:_____

Danger(s) this animal presents:

_____

_____

_____

_____

_____

_____

_____

How to identify this animal:

_____

_____

_____

_____

_____

_____

_____

Tips for avoiding or dealing with this animal:

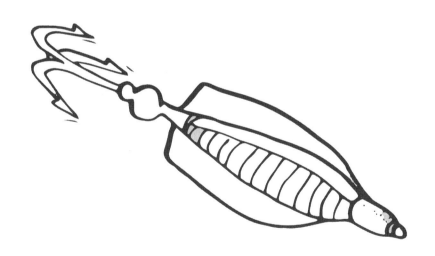

Animal #3:_____

Danger(s) this animal presents:

_____

_____

_____

_____

_____

_____

_____

How to identify this animal:

_____

_____

_____

_____

_____

_____

_____

Tips for avoiding or dealing with this animal:

Color this dangerous animal correctly.

Draw your favorite dangerous animal below!

Hunting, Trapping, and Fishing:

Use books or the internet to find five hunting tips and write them below!

Tip #1:

_____

_____

Tip #2:

_____

_____

Tip #3:

_____

_____

Tip #4:

_____

_____

Tip #5:

_____

_____

Use books or the internet to find five kinds of traps you can make from what is found in nature.

Trap Name:	Use:	Instructions:

Fishing:

Use books or the internet to research each category of fishing and answer the questions below.

Ponds and Lakes:

Three types of fish that live there:

One effective lure for each fish:

Three fishing techniques for this category of fishing

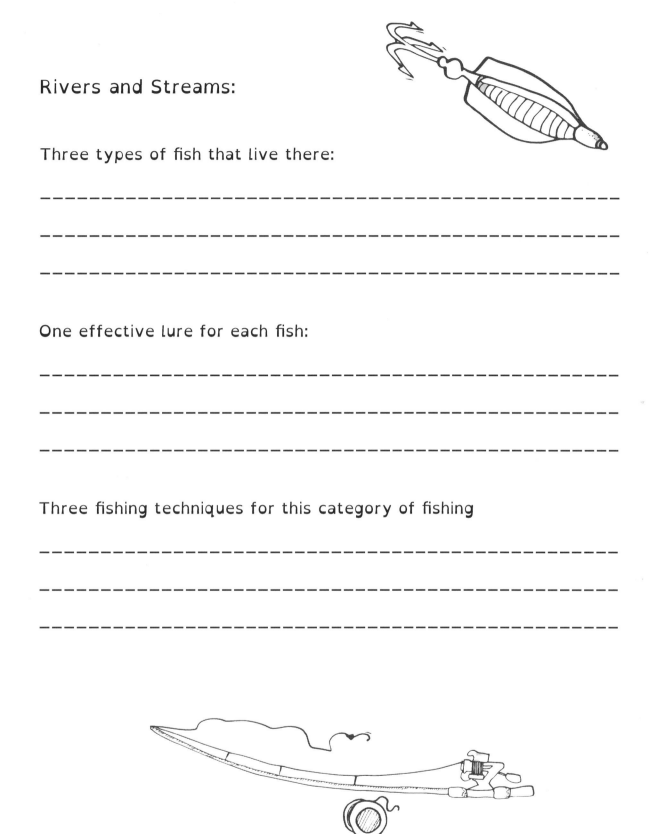

Rivers and Streams:

Three types of fish that live there:

One effective lure for each fish:

Three fishing techniques for this category of fishing

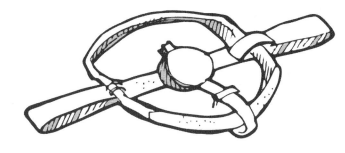

Preparation:

Pick five animals in your area and find out how to skin, clean, and prepare each animal. Write down your discoveries below

Animal #1:

_____

_____

_____

Animal #2:

_____

_____

_____

Animal #3:

_____

_____

_____

Animal #4:

_____

_____

_____

Animal #5:

_____

_____

_____

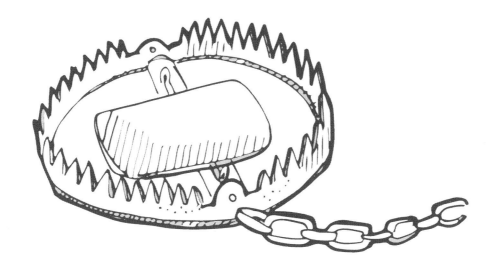

CHAPTER 2: THE THREE SURVIVAL ESSENTIALS

There are three physical things that every human needs in order to survive. These are fire, water, and shelter.

Section I: Fire

Fire Building:

Research three types of fires with different purposes and record what you find below.

Type of Fire:	Use:	Instructions:	Drawing:

Type of Fire:	Use:	Instructions:	Drawing:

Fire Maintenance:

Research three maintenance tips for each type of fire you choose and write them below.

Type of fire:_____

Tip #1:

_____

_____

Tip #2:

_____

_____

Tip #3:

_____

_____

Type of fire:_____

Tip #1:

_____

_____

Tip #2:

_____

_____

Tip #3:

_____

_____

Type of fire:_____

Tip #1:

_____

_____

Tip #2:

_____

_____

Tip #3:

_____

_____

Section II: Water

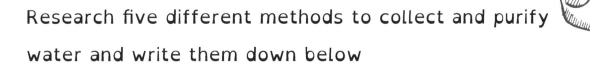

Research five different methods to collect and purify water and write them down below

Method:	Instructions:	Drawing/Diagram:

Method:	Instructions:	Drawing/Diagram:

Section III: Shelter

Use books or the internet to find three types of shelters and write your discoveries below.

Type of Shelter:	Instructions:	Draw:

Type of Shelter:	Instructions:	Draw:

Research two useful tips for each shelter and write them below.

Shelter #1:

Tip:_____

_____

_____

_____

Tip:_____

_____

_____

_____

Shelter #2:

Tip:_____

_____

_____

_____

Tip:_____

_____

_____

_____

Shelter #3:

Tip:_____

Tip:_____

Do it yourself!

With your parents' permission, try to make one of the shelters you have chosen. Draw your shelter below:

CHAPTER 3: HOW TO...

In this section, you will discover how to do many things that are very essential to survival.

Section 1: Knots and Lashings

Knots:

Use books or the internet to find ten knots and write your discoveries below.

Name of knot:	Use(s):	Instructions:

65

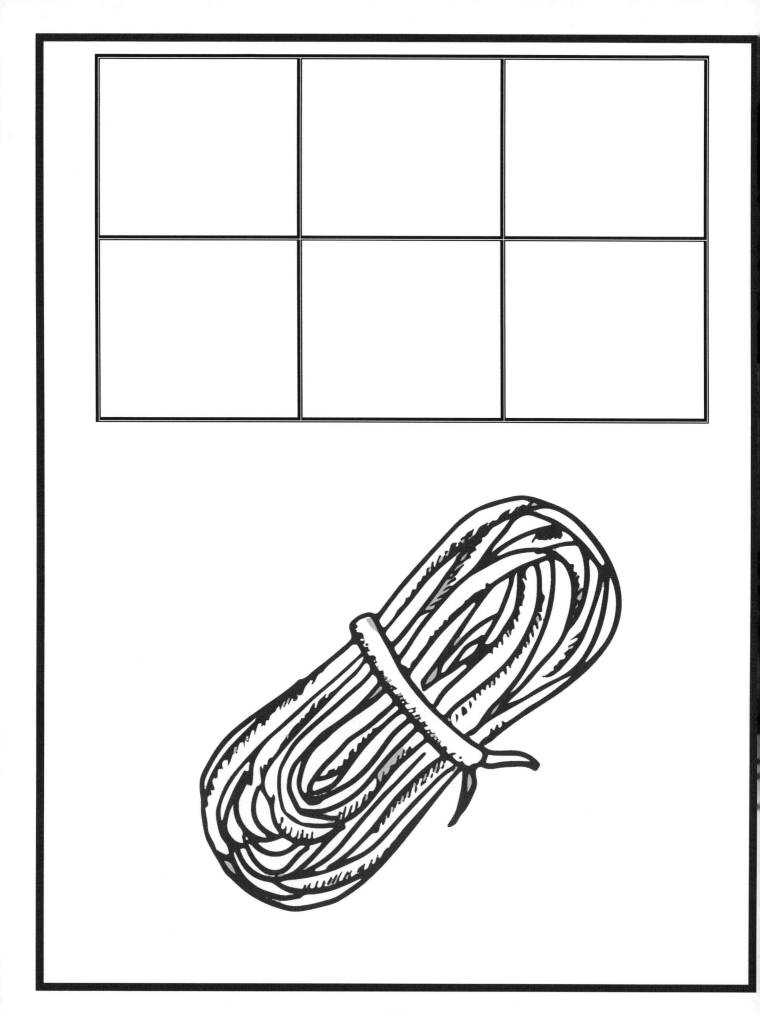

Try it yourself!

Pick three of the knots you researched and try to tie them.

Draw your knot below

Knot #1:

Knot #2:

Knot #3:

ashings:

Research three different kinds of rope lashings. Record
what you find below.

Name of Lashing:	Use:	Instructions:	Drawing:

Section 11: Orienteering

Use books or the internet to find the steps of orienteering and write them down below.

_____

_____

_____

_____

_____

_____

_____

Try it yourself!

1. Pick three bearings and write them down below.

   ----

2. One bearing at a time, find your direction of travel and take ten paces for each bearing.

   ----

3. Look at where you started and where you stopped.

Section III: First Aid

Research each of these injuries and fill out the chart below.
Make sure all information is up-to-date.

Type of Injury:	Treatment:	Maintenance:	Other Notes:
Small Cut			
Large Cut			
Minor Fracture			

Compound Fracture			
Scrape			
Bloody Nose			
First Degree Burn			

72

Second Degree Burn			
Third Degree Burn			
Dehydration			
Snake Bite			

Write or draw what you <u>think</u> should go in a first aid kit below.

Do some research and write or draw what should actually go in a first aid kit below.

_____

_____

_____

_____

_____

Section IV: Survival bag

Use books or the internet to find five tips concerning survival bags (they are also called: emergency packs, bug-out bags, disaster preparedness packs, etc.).

Tip#1:_____

_____

_____

_____

Tip#2:_____

_____

_____

_____

Tip#3:_____

_____

_____

_____

Tip#4:_____

_____

_____

_____

Tip#5:_____

_____

_____

_____

Do some research and write or draw what should go in a survival bag below.

Do it yourself!

Make a survival bag filled with all of the things you wrote or drew above.

You are now ready to face survival scenarios.

CHAPTER 4: SURVIVAL SCENARIOS

This section is full of survival scenarios that have been written out for you. Answer all of the scenarios with your recommended actions, then look in books or online to see if you were correct.

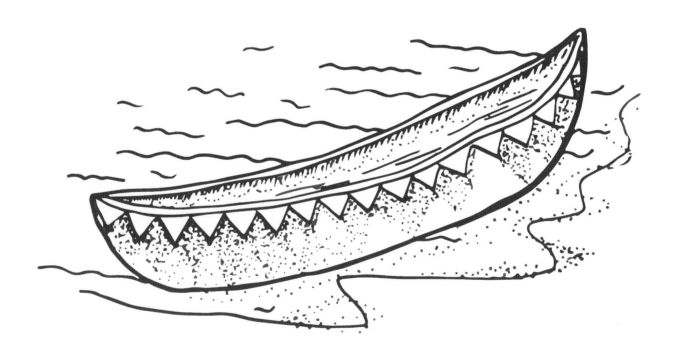

Scenario I:

You and your two friends, John and Jake, are hiking in Shades State Park. After a little while of hiking, Jake slips and falls down a ravine. After safely and slowly descending the ravine, you and John discover that Jake has broken his leg.

Notes: 1. The nearest hospital is sixty minutes away. 2. It is a compound fracture. 3. Jake has a phone.

Recommended Actions:

_____

_____

_____

_____

_____

_____

_____

Scenario II:

Part 1

You and your friend Tyler are out camping. When you go to get some firewood, you find a snake.

Notes:

1. It may or may not be venomous.

2. You are about three feet from the snake.

3. Your friend is within hearing distance.

Recommended Actions:

_____

_____

_____

_____

_____

_____

_____

Scenario II:

Part 2

You call Tyler over and tell him to stay clear of that area.

Tyler does not listen to you and tries to catch the snake.

He grabs it by the tail and it bites him.

Notes:

1. You have discovered that the snake is venomous.

2. Tyler has been bitten in the hand.

3. Tyler has a phone.

Recommended Actions:

_____

_____

_____

_____

_____

_____

_____

Scenario III:

You and your friend Tyler are backpacking in the backwoods of Kentucky. You wake up one morning and find that your camp has been ransacked by bears and that all of your food is gone.

Notes:

1. It would be a one day trip in order to get back to civilization.

2. You both have camping gear (knives, sleeping bags, a tent, etc.).

3. Neither of you have a phone.

Recommended Actions:

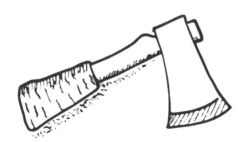

Scenario IV:

You and your two friends John and Taylor are hiking in Turkey Run State Park. After about two hours of hiking, you realize that John has stopped sweating, has started to stumble, and has begun to feel dizzy. He has also been complaining that his head hurts.

Notes:

1. It is 90°(F).

2. You are two hours away from your camp.

3. You have extra water.

Recommended Actions:

_____

_____

_____

_____

_____

_____

_____

Scenario V:

You and your friend Nathan are building a signal fire, What are some things that you do differently from a normal fire?

Answer:

Scenario VI:

You and a group of your friends are camping in a valley. It is a very rainy day and one of the adults with you gets an alert on the car radio saying that there are flash flood warnings in your area.

Notes:

1. There is a river one-hundred yards from your campsite.

2. The nearest town is thirty minutes away.

3. The adult with you has a phone.

Recommended Actions:

_____

_____

_____

_____

_____

_____

_____

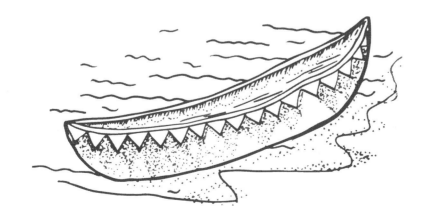

Scenario VII:

Part 1

You and a group of friends are hiking in Shakamak State Park. When you go off to filter some water from the lake to refill your water bottle, you become separated from your group.

Notes:

1. You did not walk very far to get water.

2. The lake was south of your group.

3. You do not have a phone with you.

Recommended Actions:

_____

_____

_____

_____

_____

_____

_____

Scenario VII:

Part 2

After your first attempts to find your group fail, you decide to go looking for them. After about an hour of searching, you are hopelessly lost.

Notes:

1. You know that the trail runs along the north side of the lake.

2. You have a whistle in your emergency pack.

3. You have enough food in your pack to last two days.

Recommended Actions:

_____

_____

_____

_____

_____

_____

_____

Scenario VIII:

You and your friend John are winter backpacking in Iron Mountain, MI. During a long hike, John falls into a nearby stream. By the time you get back to camp, John's body starts to shut down from hypothermia.

Notes:

1. It would be a one day trip to the nearest town.

2. Neither of you have phones.

3. You both have extra warm clothing.

Recommended Actions:

_____

_____

_____

_____

_____

_____

_____

Scenario IX:

While canoeing on the Colorado River with your friend Grant, your canoe capsizes.

Notes:

1. There is a strong current.

2. You and Grant are the only ones nearby.

3. You are at least 100 yards from the shore on either side.

Recommended Actions:

_____

_____

_____

_____

_____

_____

_____

Scenario X:

You are out hiking with your friends John and Taylor. About six miles into the hike, John starts having a stroke.

Notes:

1. You are ten miles from civilization

2. Taylor has a phone

3. The nearest trail access is six miles away

Recommended Actions:

_____

_____

_____

_____

_____

_____

_____

For Notes:

Made in the USA
Columbia, SC
16 January 2019